for beginners

Advanced Fire Starting for Beginners

Explore Various Methods for Igniting Fires in Challenging Conditions and Without Modern Tools

Printed in the United States of America

First Edition: 2025

This copyright page includes the necessary copyright notice, permissions request information, acknowledgments for the cover design, interior design, and editing, as well as the details of the first edition.

www.littlebiggiant.com

Introduction

Wind cut through the trees. Night fell faster than the map said it would. A small figure crouched over a pile of damp leaves. Fingers trembled. Every match had already been used. This was the moment all the practice drills had been about. The plan was simple: get an ember, grow it into a coal, feed it. The plan felt impossible with rain leaking down a sleeve. "This won't catch," came a whisper. Breath hitched. The thought surfaced sharp and cold: what if this fails?

A known fact: archaeologists have found evidence that people controlled fire hundreds of thousands of years ago. That means the skills about to be tried here are not tricks. They are ancient tools passed down by practice and patience. A bow drill makes an ember by friction. A piece of flint and a steel striker can shoot tiny, hot sparks. A lens can focus sunlight into tinder on a bright day. None of those things are magical. They are methods. The hands now tried them one by one. A loosened bow rubbed a spindle until the wood smoked. A flint sent a sharp shower of sparks that landed on a tiny rag of dry plant fibers. For a moment, nothing. Then a gray thread of smoke curled up. Relief warmed the chest like the first breath after being underwater. Still, the ember was fragile. It could die with one wrong move, or it could grow and change everything about the night. Which would it be? The next breath felt like a decision.

Table of Contents

Chapter 1

Foundations of Primitive Firecraft

By the end of this chapter, you will understand how fires start through basic combustion and oxygen control, you will learn how to choose and prepare fuels, kindling, and tinder, and you will know how to stay safe and responsible around fire.

Foundations of Primitive Firecraft: Understanding combustion and oxygen control

Fire is a visible sign of a chemical process that needs three ingredients: fuel, heat, and oxygen. Remove any one of these parts, and the flame goes out. In practical terms, understanding combustion means recognizing that wood, coal, and other fuels will burn only when they can reach and maintain enough heat, and when enough air can reach the flames to feed the reaction.

Air is not a neutral helper. It moves and flows, and its movement changes how a flame behaves. When you open a vent or lift a covering, air rushes in and can make a flame grow taller and brighter. When air is blocked or when a flame is smothered, heat builds up and the fire may sigh and then die. This is why most fires need a craft of airflow: a gentle, controlled supply of oxygen keeps the flame steady without blowing it into a dangerous, roaring blaze.

Different fuels burn in different ways. Light, dry materials catch quickly and burn hot but short; heavier fuels burn longer but may require more heat and better airflow to start. Damp wood, leafy debris, or wood with a lot of sap or resins tends to hiss, steam, or smoke and can be hard to keep burning. Dryness matters more than size alone. In general, the drier the fuel, the more predictable the burn, provided there is enough heat and airflow to start it.

Two quick mental models help you think about oxygen control. First, airflow acts like a lever: small changes in how air moves through the fire area can make a big difference in how quickly a flame catches and how steadily it burns. Second, oxygen isn't a single blast of air; it's a continuous stream that you manage with the shape of the fire pit, the placement of fuel, and the moment you introduce or reduce airflow. With those ideas, you can craft a fire that glows reliably rather than sputters.

Example 1: On a calm, dry hilltop, you notice a small pile of twigs mixed with thin bark and very dry pine needles. The needles catch fast but burn briefly. You arrange the tinder and kindling in a conical mound that leaves space for air to flow up through the center. When you expose a thin spark or a small flame to the tinder, a slight puff of warm air from the surrounding air helps the tinder catch. The flame grows smoothly as you feed it with light sticks, and you keep a steady rhythm of airflow by gently fanning with a hand-sized shield of dry leaves or a flat stone to direct air to the flame without blowing it out.

Example 2: In a damp forest edge with recent rain, you find dry pockets under a fallen log and a small, sunlit spot where wood is dry enough to catch. You choose tinder from dry grasses and bark fibers, and you split some thin logs to expose dry interior wood. The fire's initial heat comes from a compact bundle of dry tinder and small twigs; you carefully arrange larger sticks around the edge so air can circulate along the sides. Because the moisture in bigger pieces slows ignition, you time your feeding so you don't smother the flame—adding fuel gradually and watching how heat builds and airflow shifts the flame's shape.

Practical note: when you're thinking about combustion and oxygen, keep safety front and center. A steady flame that is easy to control is safer than a large, fast flame that can get out of hand. If you ever feel unsure, pause, reassess the location and the amount of fuel you've placed, and remember that wind, nearby vegetation, or a sloped terrain can dramatically influence how a fire behaves.

Selecting and preparing fuels, kindling, and tinder

Fuel, kindling, and tinder are three distinct parts of the fire-creation toolkit. Each plays a specific role in the ignition sequence, and understanding their differences helps you choose wisely and prepare effectively.

Fuel means the larger pieces that feed the fire once it's lit. Good fuels are dry and have a stable burn. They come in a range of densities: softwoods like pine or spruce ignite easily and burn quickly; hardwoods like oak or beech take longer to catch and burn more slowly and steadily. The key is dryness. Freshly cut wood often has moisture that slows ignition and can smoke rather than burn. If your goal is a reliable start, seek wood with a dry feel and a light, hollow-sounding strike when tapped. Larger fuel pieces should be placed to encourage air to circulate around them, which helps heat reach the inner surfaces and keep the fire burning.

Kindling is the next tier. It includes small sticks, slivers, and wood scraps that ignite easily and carry the fire from tinder into larger fuel. Kindling should be thin enough to catch from the tinder's heat but sturdy enough to hold shape as it starts to burn. Fine shavings or small split twigs often work well. The trick with kindling is to arrange it so air can move through and around it, maintaining a steady flame that will readily ignite the larger fuel pieces.

Tinder is the spark-bridge material—the stuff that catches fire from a small spark, ember, or flame and then feeds it into a larger flame. Tinder is typically easy to ignite and has a fibrous, fluffy, or highly absorbent structure. Common tinder materials include dry grasses, bark fibers, shredded paper, dried leaves, and tiny dry wood shavings. The goal is to pick tinder that lights quickly and then holds on long enough for the kindling to start burning with enough heat to ignite the larger fuel.

A practical approach to selecting and preparing materials is to prioritize dryness and arrangement. Look for dry pockets in the environment, test pieces by bending or snapping them—dry wood snaps crisply; damp wood bends and may feel soft inside. Split larger logs to expose dry core wood, remove obvious moisture bands, and store gathered materials in a dry place or cover them from rain if you have time. Maintain a simple rule: the drier your tinder and kindling relative to your fuel, the more likely your fire will catch and stay lit.

Two micro-cases demonstrate this approach in action.

Micro-case A: You're in a sunny, dry yard. You gather a handful of pine needles (tinder), a few small twigs (kindling), and thin branches (fuel). You test each material's dryness by bending and twisting; needles crackle but don't bend, indicating dryness. You arrange tinder in a loose nest, with a space at the center for a spark. You add a few shavings of bark and small sticks around the tinder, ensuring air can move through. When ignition happens, you feed the flame with gradually larger pieces, watching as the fire grows without rushing heat into the larger logs.

Micro-case B: You're at a damp riverside with damp ground. You search beneath logs and in sheltered pockets for dry strips of bark and tiny dry twigs that have escaped the moisture. You collect these as tinder and make sure the flame can latch onto the smallest pieces first. You split a couple of larger logs to reach dry wood inside and place them to create a bed where airflow can pass up through the kindling and tinder. The tinder catches more slowly, but once it does, the flame climbs steadily into the tinder and then onto the kindling and fuel.

Safety note: always prioritize safety when selecting and preparing materials. If you cannot find dry tinder, or if the surrounding area is very damp or windy, reconsider trying to ignite a fire. Dry, stable conditions are safer and more reliable for learning and practicing.

Fire safety, risk assessment, and responsibility

Fire safety is not a single action; it's a continuous practice of assessment, preparation, and control. You can reduce risk by planning ahead and staying attentive to your environment.

Begin with location and wind. Choose a bare, clear area away from leaves, brush, tents, or structures. Make sure there is nothing that could catch fire nearby, and that the ground beneath the fire is flat and nonflammable. Consider the wind direction and speed. If the wind picks up suddenly or if a gust could blow embers toward dry areas, pause or move to a safer spot.

Create a safety boundary. A simple circle or ring of cleared ground provides a space where embers won't jump to dry materials around you. Have a ready source of extinguishing material—water, dirt, or sand—and be prepared to smother the fire if needed. Do not operate a fire close to people, pets, or flammable materials, and always keep a plan for quickly putting out the fire if anything goes wrong.

Legal and environmental responsibility matters. Check local rules about burning in your area, especially during dry seasons when fires are more dangerous. Be mindful of wildlife, plants, and ecosystems, and leave no trace when you're done. When you leave, make sure the fire is completely extinguished and cold to the touch. If you're unsure about the air quality or the safety of continuing, step away and reassess before you proceed.

A simple risk assessment framework can help you make quick decisions. Identify the hazard (for example, wind gusts, dry grass nearby), consider how likely it is to occur and how severe it would be, and then decide on a mitigating action (such as extinguishing the flame or moving to a protected area). This mental checklist can save time and reduce danger in the moment.

Two brief examples of applying safety and responsibility:

Example 1: You've built a small, contained fire in a prepared circle on a calm day. The flame grows, but you notice a stray ember drifting toward dry grass. You pause, lower the flame by reducing airflow, and gently smother the embers with dirt. After a minute, you re-check the area and confirm there are no hissing sounds, no smoke, and no heat through the soil around the circle. You proceed to fully extinguish the remaining flame, then clean up the site.

Example 2: You are camping in a windy area with scattered pine needles around the fire site. You build a low pit with stones, place the tinder and kindling in a sheltered setup that blocks direct wind, and watch for any signs of embers escaping. If wind picks up too strongly, you stop and relocate to a more protected area or wait until conditions calm. When you leave, you ensure the fire is completely out and the site is clean.

Key Takeaway: Comfort with fire comes from knowing how fuel, heat, and air work together, and from practicing safety and responsibility every time you approach ignition.

Coming up next: Friction Methods — Hand-Drill and Bow-Drill

1) Choosing compatible wood species and shaping components

2) Constructing a reliable hearth, socket, and spindle

3) Technique, rhythm, and troubleshooting for consistent ember creation

Coming up next

Friction Methods — Hand-Drill and Bow-Drill

1) Choosing compatible wood species and shaping components

2) Constructing a reliable hearth, socket, and spindle

3) Technique, rhythm, and troubleshooting for consistent ember creation

Chapter 2

Friction Methods — Hand-Drill and Bow-Drill

You start by picking pieces that will let friction do the work without fighting you. Dry wood is easier to rub than soaked wood, so look for pieces that feel light and crack a little when you squeeze them. Avoid wood with obvious moisture, mold, or resin pockets, because they can steam instead of heat up cleanly. The goal is for the two parts you spin to “grab” each other little by little, not slide away or crumble.

Think about matching the general character of the pieces. A spindle and a hearth board that come from wood with similar hardness tend to rub better together. If you test two candidates by rubbing the tip of one against the surface of the other, you want to feel steady friction that doesn’t grind to a stop immediately or slip away too easily. The ends of the spindle should be smooth and rounded; a rounded tip reduces hot spots and keeps the contact area comfortable for your hands. The hearth board should have a small, shallow groove or depression near the center to hold the dust and char that form as you work. A light notch at the edge helps trap material that could otherwise fly away as you spin.

Step 1: Choose dry, straight-grained pieces. Look for wood that feels firm but not brittle. If you can connect two pieces so they rub cleanly without snagging, you're in the right ballpark. When in doubt, pick a pair that you can test quickly with a simple rubbing test: press the spindle tip and move it across the hearth. If it glides with steady resistance and doesn't slice or tear, that's a good sign.

Step 2: Pair compatible pieces and shape them. Aim for a spindle that isn't too thick or too thin for your hands, with ends rounded to reduce friction hot spots. Shape the hearth board so there's a shallow bowl or groove where the dust can collect. The notch near the edge should be just large enough to catch ember dust without letting it spill away. If you plan to practice with a bow-drill later, leave a little extra surface on the hearth so you can adjust as you go.

Step 3: Prepare for rubbing by testing a dry run. Take a dry pair and rub gently for a short period. If you can feel a growl of heat and see a tiny puff of dust, you're near the zone where embers begin. If there's smoke but no heat, your pieces may be too dry and brittle in a bad way, or you may need a bit more moisture control (gentle air exposure, not soaking). If the pieces slip or chatter, re-check the fit and smoothness of the rounded ends and the groove in the hearth. Small adjustments here save big frustration later.

Example 1: Hand-drill setup with two dry pieces. You pick a straight-grained spindle and a compatible hearth board. You shape the spindle ends to a smooth point and the hearth bowl to catch dust. You position the board on a firm surface, place the spindle in the notch, and press your hands together. You begin rubbing slowly at first, then gradually increase speed while maintaining consistent pressure. The dust swirls in the groove, heat builds, and after a careful push-and-roll, a tiny ember glows in the pocket of dust. You tamp the ember into a loose pile of tinder and blow gently to grow the flame.

Example 2: Bow-drill pairing with simple socket. You choose two pieces that feel evenly dense and smooth, and you shape a slender spindle while keeping the hearth board flat. You wrap a cord around a flexible bow, loop a string, and set up a small socket or guard to rest the spindle against your hand. You move the bow in a steady, even rhythm, letting friction create heat where the dust gathers. After several seconds of consistent motion, a warm glow appears in the dust nest. You pause, breathe, and gently coax that glow into a bright ember by adding a pinch of tinder and a careful puff of air.

Constructing a reliable hearth, socket, and spindle

A solid hearth, a steady spindle, and a safe hold are the backbone of friction lighting. If one piece is loose or unstable, the ember won't form, or the heat will wear you out before you get a spark.

Step 1: Build a stable hearth surface. The hearth board should be flat and hard enough to resist gouging as you rub. Place it on a stable, level surface. A steady base makes every stroke more reliable. The groove or notch helps trap friction dust, which is the key to transferring heat into the dust pile that becomes the ember.

Step 2: Create a snug socket or hand guard. If you're using a hand-drill, a small socket or a simple guard protects your fingers from the heat and helps you apply even pressure. A stone, a smooth rock, or a carved wooden cup can serve as a pocket for the spindle's base where your palm sits. The socket shouldn't wobble; a steady hand lets you keep the right angle and motion.

Step 3: Shape and balance the spindle. The spindle should feel comfortable to hold and long enough to allow full rotation without hitting your hand. Turn the ends smooth so they don't snag in the wood, and ensure the midsection is straight and true so that rubbing creates uniform heat along the contact point. If you notice wobble, re-check the ends and the straightness, then smooth out any rough patches.

Step 4: Prepare the tinder nest and ember pathway. Have a small pile of tinder within easy reach, plus a waiting bed of fine dust or sawdust from your own hearth work. The ember will form in the tiny pocket you carved into the dust; you want a gentle, ready environment for that moment when heat hits dust and the ember lights.

Technique, rhythm, and troubleshooting for consistent ember creation

The heart of this skill is a steady rhythm, correct pressure, and calm adjustments. When you can keep your hands aligned and your breath steady, you'll create embers more reliably.

Step 1: Find a comfortable rhythm. Start slow, then increase speed gradually while keeping consistent pressure. Think of your hands as coils in a gentle engine: the top hand guides the motion, the bottom hand maintains the pressure, and both move together in a smooth, continuous cycle. If your hands shove or slip, slow down and reset.

Step 2: Keep your angle and contact stable. The spindle should rub in a straight line along the hearth's groove or surface, not at an angle that digs into wood. If the line veers, adjust your wrist and elbow so the motion is even and predictable. A consistent contact point matters more than speed.

Step 3: Watch for signs of ember readiness. You'll see a dark dust ring forming and a subtle glow in the dust pocket. The ember often looks like a tiny coal or a warm, orange speck. When you notice that glow, stop rubbing and gently blow toward it to feed it into a flame.

Step 4: Troubleshooting quick fixes. If there's no heat after a long rubbing session, check moisture first. If the wood feels damp or cool, give the pieces a few minutes to acclimate to the air, or switch to drier pieces if available. If the dust stays dusty and refuses to glow, you might need more consistent pressure or a slightly better fit between spindle and hearth. If the ember collapses when you try to fan it into flame, back off the airflow and give the ember more time to catch tinder. If the spindle wobbles, recenter the ends, re-check the notch, and ensure the hearth isn't slipping.

Two micro-cases show how the same principles feel in action:

Micro-case A: A quiet practice session with hand-drill. You test two new pieces gathered from a dried branch and a small wooden plank. You shape the spindle ends, notch the hearth, and try a calm, deliberate rhythm. Heat builds in short bursts, and you see a faint dust swirl as the ember forms. You transfer it to tinder and breathe softly to catch the flame. It works cleanly, and you feel confident you could repeat it with similar pieces.

Micro-case B: A windy afternoon with bow-drill practice. You set up a simple socket and a bow with moderate tension. You copy the same steady rhythm, but the motion is faster because the bow provides mechanical advantage. In a few moments, the dust glows, and the ember appears in the pocket. You lift your tempo slightly, pack tinder, and coax the flame to life.

Key Takeaway: Consistent ember creation comes from choosing dry, compatible wood, shaping each piece for smooth rubbing, and practicing a steady, controlled rhythm that keeps heat building and embers forming.

Coming up next Subtittle: Advanced Friction – Pump Drill and Fire Plow

1) Leveraging mechanical advantage for reduced effort,

2) Crafting durable parts with minimal tools,

3) Adapting methods for soft, wet, or fibrous woods

Chapter 3

Advanced Friction — Pump Drill and Fire Plow

By the end of this chapter you will be able to leverage pump drills to reduce effort, craft durable friction parts with minimal tools, and adapt the pump-drill and fire-plow methods for soft, wet, or fibrous woods.

Leveraging mechanical advantage for reduced effort

Step 1: Build a simple pump drill using a spindle, a bearing point, a drive cord, and a downward push handle. The goal is a smooth circle from your wrist, not a fast shove. A standard setup uses a straight stick as the spindle, a short notch in a wooden bearing, and a cord wound around the spindle that runs through the bearing toward a downward handle or weight. The weight or push-handles give you leverage so your arm doesn't have to spin the spindle all at once. The cord turns the vertical motion into rapid rotational motion, so you can keep a steady rhythm without burning out your shoulders. The rationale is simple: mechanical advantage lets you convert a bigger input movement into a quicker, controlled spin, which raises the odds of slipping less and keeping heat focused where the tinder lies.

Step 2: Position your hearth board and make sure the spindle sits securely in its socket. A solid seating point reduces wobble and keeps friction heat concentrated on the same spot. Line up the spindle so the point cuts evenly across a dry notch or a small groove you carved in the hearth. Your goal is a steady, almost musical rhythm: consistent pressure, consistent speed, and little side-to-side jostling. The rationale here is that stability matters as much as speed; a stable contact area concentrates heat and creates the fine dust that turns into ember matter, rather than scattering heat across wood fibers.

Step 3: Practice a comfortable, repeatable motion with moderate speed and even pressure. Don't press too hard or you'll stall; don't hurry and risk a wobble that blows out the heat. Keep your wrists relaxed, your elbows close to your body, and your gaze focused on the point of contact. The rationale is control: a consistent motion builds enough heat in the right place to spark dust and paves the way for a usable ember. With practice, you'll notice the dust pile growing lighter and finer, almost glowing before you see it as a small ember.

Example: A short, real-world micro-case shows how this works. A camper named Niko sets up a pump drill with a dry cedar hearth and a straight spindle. He sits on the ground, steadying the board in his lap, and begins a slow, even rhythm. Within about a minute, a small, glowing speck appears in the dust—an ember that grows when he blows gently on it and slips it into a nest of dry pine shavings. The ember rests there briefly, then he moves it into his tinder bundle and lights a tiny flame. The result is a controlled, repeatable process rather than a frantic scramble for dry wood.

If you don't see progress, check these quick fixes: adjust the cord tension so the cord doesn't slip on the spindle; smooth the bearing surface so the spindle turns more freely; ensure the hearth board isn't damp or glossy, which can steal heat away; and confirm the wood you're using is dry enough for friction to work. When you can keep the rotation steady and the ember formed, you'll feel the method click.

Crafting durable parts with minimal tools

Step 1: Choose durable, simple components that you can shape with basic tools. Pick a hardwood for the spindle that resists wear and a sturdy, flat piece for the hearth board. You don't need fancy tools—just a sharp dull knife, a small saw if you have one, and a willingness to test fit. The rationale is practical: tough, reliable parts last longer under friction, reduce the chance of piece failure mid-use, and give you a steadier progression from friction to flame.

Step 2: Shape and fit the spindle, socket, and bearing so they work smoothly together. Use shallow, clean notches or grooves to guide the cord and to keep the spindle aligned. Carve a small socket or notch in the hearth board to cradle the spindle tip so it won't slide out during use. The rationale is precise fit: smooth interfaces lower friction where you don't want friction, and they keep the spinning action focused on heat creation rather than on fighting parts.

Step 3: Test and refine with light trials before you go for a full ember session. Do a few quick spins, watch the smoke, and listen for a steady "hiss" as the dust forms. If you hear a grinding sound or see the spindle wobble, pause, re-align, and adjust the notch or bearing. The rationale is to catch misfit early; a small adjustment now saves failed attempts later and builds reliable muscle memory for these tools.

Example: In a backyard test, Kai uses a simple wood spindle with a rounded tip and a loose-earth hearth board. He sculpts a narrow socket in the board and lines up a straight groove for the cord. After a couple of short practice runs, he achieves a clean groove and stable rotation. On the next attempt, the friction dust gathers neatly, forming a visible ember after a few more passes. He places the ember into a bed of dry tinder and lights a small flame. The process feels efficient because the parts cooperate consistently.

If parts wear or don't hold their shape, swap or re-carve the components using the same simple tools. Don't overcomplicate the design—keep it simple, sturdy, and easy to replace. A durable setup is a dependable tool you can rely on when conditions are less than perfect.

Adapting methods for soft, wet, or fibrous woods

Step 1: Assess the wood you're working with before you start. Soft, wet, or fibrous woods don't hold heat as well and can shed after only a few seconds of friction. If you have a choice, look for drier, denser fibers or inner core wood, which burn longer and resist tearing apart too quickly. The rationale is to work with the material's natural tendencies; dry, compact wood gives you better heat retention and a steadier ember.

Step 2: Adjust your technique to the wood's real limits. If the wood is fibrous or damp, you may need to slow your pace a touch, apply more consistent pressure where the grain runs, and keep your contact point steady longer to coax dust from tougher strands. The rationale is to compensate for water and fiber properties by maintaining heat in a single spot rather than scattering it across uneven surfaces, which helps you form embers even in less-than-ideal wood.

Step 3: If you can’t find dry wood, use a two-step approach. Start friction on dry inner material or bark-free sections where possible, then shift your effort toward a drier tinder bundle so that the ember can be transferred quickly into flame. The rationale is to keep ember formation possible even when material quality isn’t perfect, by isolating the hot spot from damp, fibrous surfaces and moving quickly to ignition.

Example: A scout team is in a damp clearing with wet birch. They choose a dry innerboard section for the hearth and a cedar spindle, then slow the pump drill rhythm to keep friction heat concentrated. They notice the dust forms more slowly, but after a few careful passes, a small ember appears. They nest the ember into a dry tinder ball and gently blow, producing a modest flame. The team records the outcome as a successful start despite the moisture in the wood.

Troubleshooting and quick fixes

1) Ember fades or never forms: verify wood dryness, reduce moisture in the work area, and

ensure the ember is placed into dry tinder immediately after formation.

2) Spindle slips or wobbles: tighten the seating in the bearing, re-check the socket fit, and make sure the cord is taut but not stretched out.

3) Cord wear or fraying: replace the drive cord with a fresh length and inspect the winding path for friction hotspots.

4) Heat sneaks into unintended surfaces: ensure the contact area is narrow and focused, not broad, to prevent heat loss across wood grains.

5) Parts break or crack under pressure: carve new parts from solid wood rather than forcing damaged pieces back into service; keep spare spares on hand when possible.

Templates and quick-tools you can copy

- Pump Drill Parts Checklist: spindle, bearing point, drive cord, hearth board, socket or bearing surface, downward handle or weight, dry tinder for testing, spare cord.

- Quick-fit Breathing Guide: a one-page reminder for maintaining a steady rhythm, a simple way to test for ember formation, and a note on moving embers into tinder.

Key Takeaway: Durable, simple parts and steady, controlled motion turn friction into ember more reliably, even when wood isn't perfect.

Coming up next: Spark-Based Ignition — Flint, Steel, and Ferrocerium

You'll learn how to generate focused sparks and position them effectively, prepare char cloth, punk, and alternative tinder bundles, and care for strikers while improvising when conventional tools are scarce. After practicing the pump drill and fire plow, you'll have a dependable friction-based foundation that supports the next, faster ignition methods. In the next chapter, you'll explore how small, hot sparks can ignite tinder quickly and under a wider range of conditions, bringing your fireplace building skills to a new level.

Chapter 4

Spark-Based Ignition — Flint, Steel, and Ferrocerium

By the end of this chapter, you will be able to generate focused sparks with flint, steel, and a ferrocerium rod and use them to light tinder reliably in common conditions.

1. Generating focused sparks and positioning them effectively

When you strike metal against stone or a ferro rod, you create tiny, hot sparks. Those sparks are not flames themselves; they are tinder carriers that start a small, hot ember if they land on the right material. The trick is to control two things at once: the surface you strike, and the place you aim the sparks.

First, choose your surface. With traditional fire steels, you strike the steel edge along a hard, sharp-edged rock or a properly sharpened striker. With a ferrocerium rod, you strike the rod with a hardened steel striker. The goal is to throw a shower of bright sparks rather than a slow, dull flare. Practice helps you feel the rhythm: a quick, firm scrape or strike tends to produce more consistent sparks than a slow or glancing motion. The angle matters too. Aim for a forty-five-degree tilt, not flat, and keep the motion short and sharp rather than a long, lazy pull. The contact area should be the edge or a small ridge on the striker, not a broad, dull surface.

Now, position your tinder correctly. In the mouth of your fire setup, you want a tiny, fluffy nest of tinder ready to catch. That nest can be char cloth, dry lint, fine wood fibers, or dry grasses. The key is fluff and air. If the tinder is compact, the sparks will bounce off or fail to light it. If it's loose enough to hold a little heat but dense enough to keep the ember in one spot, you'll have a longer window to catch and nurture the flame.

A few practical tips to move from sparks to flame. Keep a stash of ready tinder where you can see it and reach it without fumbling. When you strike, aim toward the center of the tinder nest. If a spark lands and dies out, don't wipe it away; keep the nest quiet and patient. You'll often get a small, glowing dot—a hot ember—that you can nurture into a flame with a guided breath and a gentle puff of air.

Two short examples show how this works in different setups. Case A: You have a small ferro rod and a sturdy steel striker. You already prepared a dry, fluffy tinder nest of dry grasses. You strike in a quick, firm stroke. A bright spark lands on the grasses and glows for a moment. The ember grows as you shelter it with your hand and feed it tiny amounts of tinder until it catches fire. Case B: You're using a traditional flint and steel kit. The piece of flint is sharp and dry, and your steel edge is clean. You strike steadily, the sparks forming a shower that lands directly on char cloth. The char cloth catches quickly, turning into a bright, warm ember you blow into a steady flame.

Common problems show up, and they're usually easy fixes. If sparks are weak or sparse, check moisture in your tinder and your hands. A damp day means you'll need drier tinder and drier hands, and you may need to adjust your strike angle to find where the sparks land best. If the tinder doesn't catch after several tries, try a different tinder nest for variety—char cloth often catches where lint won't, and fine dry grass or pine needles catch where cotton fibers might be stubborn. Remember, patience matters. Sparks are a tool, not a miracle; you still need proper tinder, air, and a steady, confident motion.

Two mini-experiments to try this week. First, practice with a ferro rod and a steel striker. Make a small pile of three tinder options: char cloth, dry lint, and dry grass. Strike each surface five times, aiming at each tinder type. Note which tinder catches the ember fastest and how long it takes to grow into a flame. Second, switch to flint and steel. Repeat the same process with fresh tinder. Compare how long you wait for the ember to take hold and how much air you need to coax a flame. The goal isn't to conquer every method in one sitting but to notice which combination feels most reliable to you in your conditions.

2. Preparing char cloth, punk, and alternative tinder bundles

Char cloth is a tiny, specialized tinder that lights easily when hot sparks touch it. It is made by partially burning fabric in a low-oxygen environment, which changes the fibers into a black, leathery sheet that catches a spark like a tiny coal. If you don't have char cloth, you can use other tinder that behaves similarly under a spark: cotton lint from a dryer, shredded cotton from old t-shirts, or fine plant fibers such as dry grass or moss. The important part is that the material is dry, fluffy, and able to hold a spark long enough for you to breathe life into it.

Punk is another classic tinder. It is dry, fibrous material gathered from the inside of dead wood, tree bark, or plant fibers. Punk acts like a tiny wick: it soaks a little heat and holds a spark until you blow and feed it with more tinder. If you don't have traditional punk, you can use dry, fibrous plant matter, shredded bark, or even scraps of dry paper with caution. The trick is texture: you want something that fluffs up easily and doesn't crush into a flat pancake when you blow on it.

Creating an effective tinder bundle is a small art. You want a nest that can capture a spark and hold it long enough for your breath to turn it into a flame. A simple bundle can look like a tiny ball or horn-shaped nest of fluffy material, lightly pressed so it doesn't crumble into dust. The outer layer should be a bit stiffer so it can protect the inner ember from a gust of wind. If you're using char cloth, you place it at the heart of the nest; if you're using lint or plant fibers, you tuck them around the ember to keep heat focused.

Two practical examples help show how to assemble bundles. Example 1 uses char cloth combined with a few strands of dry grass. You cut a piece of char cloth the size of a small coin and lay it in the center of a loose nest of dry grass. The grass acts as a fuel that increases airflow and helps snap the ember into a flame when you blow. Example 2 uses silk-like plant fiber and a tiny piece of punk made from dry inner bark. The punk acts as a wick, holding the initial heat while the outer fibers feed the flame as you increase oxygen with a gentle breath.

Alternative bundles can be formed from several common, safe materials. Dry pine needles work well because they're light and airy, and dry grasses provide good airflow. Cotton lint from old clothes or towels is very common and catches sparks reliably if it's genuinely dry. Avoid damp or moldy plant matter, as moisture short-circuits the ember and defeats the spark. If you're in a place where you can't find fine tinder, practice with several options to learn which materials light easiest given your environment.

What to do if your tinder fails to catch. First, confirm the tinder is dry. A tiny amount of moisture on the fibers can prevent an ember from taking hold. Second, check the nest's structure. If it's too compact, it won't allow air to mix with the ember; if it's too loose, the ember may vanish too quickly.

Adjust the bundle by adding or rearranging fibers so the ember sits in the middle, surrounded by an air pocket. Finally, try switching tinder types. Char cloth tends to be forgiving; lint often lights quickly if dry; pine needles are resilient in windy conditions. Practice with all these options so you know what to reach for when time is tight.

Two short micro-cases to watch. Case A: You have char cloth and a small amount of dry grass. You form a compact nest with char cloth at the center and loosely surround it with grass. You strike the ferro rod with a steel striker; a spark lands, char cloth glows, and the nest flares into flame after a few breaths. Case B: You only have dry pine needles and a chunk of punk from a log. You create a shallow, layered nest with punk at the core. You strike, and the first spark lights the punk; you feed in more needles as the ember grows into a healthy flame. These cases show how different tinder bundles can be adapted to the same ignition method.

Caring for strikers and improvising when conventional tools are absent. Strikers—whether metal strips, knife blades, or other hardened edges—need some basic care. After use, wipe away oil, dirt, and moisture. Keep the edge clean and dry; moisture can dull the edge over time and reduce spark production. Inspect the striker for chips or cracks. A damaged edge can still work, but it won't perform as reliably. If you lose your proper striker, look for a hard, sharp edge nearby. The principle is simple: you need a surface hard enough to abrade or shave a tiny amount of ferrocerium or steel to release a spark. In an emergency, you can improvise with a small, sharp piece of metal from a safe object, like a clean, sturdy blade edge or a nail head with care to avoid injury. Always prioritize safety: keep fingers away from the striking line and wear eye protection if possible.

Two quick improvisation scenarios. Scenario 1 places you in a rain shadow under a tree with no ferro rod, only a pocketknife blade and a small rock. You strike the rock with the knife edge to throw a shower of sparks onto a dry char cloth. The embers glow and you breathe life into the nest. Scenario 2 uses a steel mug edge and a leftover piece of metal from a tool belt. You strike deliberately, sending sparks toward a nest of dry lint. The tiny ember catches and grows into a flame. These scenarios illustrate how you can adapt when your preferred tools aren't available, while still using the same core ideas: create heat, protect the ember, feed with suitable tinder, and coax a flame.

Key Takeaway: Sparks light the flame, but the tinder and technique carry the spark into a real fire.

Coming up next Subtittle: Concentrated Heat – Lenses and Controlled Reactions

1) Using natural lenses and fashioned magnifiers safely ,

2) Understanding and applying benign exothermic reactions,

3) Selecting scenarios where concentrated heat outperforms other methods.

Chapter 5

Concentrated Heat — Lenses and Controlled Reactions

You will be able to start a fire using lenses and controlled heat when sun is available and other methods aren't practical.

Using natural lenses and fashioned magnifiers safely

Step 1. Gather your materials: a clear, clean tinder nest and a lens you can hold steadily, like a magnifying glass or a transparent plastic bottle filled with water. Rationale: a smooth, dry tinder nest catches heat more reliably, and a clear lens concentrates sunlight into a small, hot spot that can ignite it.

Step 2. Choose a focal point. Hold the lens between you and the sun and move it until the sunlight focuses into the smallest, brightest dot you can see on the tinder. Rationale: the brightest, smallest patch is where heat is most intense, increasing your chances of ignition without burning through the tinder too fast.

Step 3. Set a stable setup. Keep your elbow steady and position the tinder at the focused spot without wiggling the lens. Rationale: movement breaks the concentration and slows or stops the ember from forming. A steady hand and a fixed angle help you build a reliable ember more quickly.

Step 4. Control the heat. Maintain focus on a tiny patch of dry tinder until you see whirls of heat or a faint smoke curl. Rationale: heat builds at the focal point; watching for subtle signs helps you know when to ease up or adjust rather than overcooking the tinder.

Step 5. Nurture the ember safely. When a small ember appears, lower the heat and gently blow or fan from the side to feed the flame, then place it into a prepared fire bed. Rationale: embers need oxygen and a safe path to a real fire; moving too fast or too much air can extinguish them.

Step 6. Safety and awareness. Never point a lens at anyone's eyes, and keep bystanders well away. Avoid aiming at reflective surfaces that could redirect light into your skin or eyes. Rationale: lenses can burn and cause serious eye damage or skin injuries in seconds; clear space prevents accidents.

Example A. On a bright day during a hike, you spot a dry patch of tinder under a small leaf-lined nest. You set your magnifier so the sun's rays converge onto the tinder. After a few careful passes, smoke lifts, and the nest catches into a flame you can feed with small sticks. The flame grows, and you move it to your fire bed, ready to build a proper fire.

Example B. You don't have a glass magnifier, but you do have a clean plastic bottle. You fill it with water, hold it steadily, and position it so the sun focuses on a pile of dry tinder. The bottle acts as a crude lens; after a short wait, a pinpoint of heat appears and sparks a tiny ember that you coax into a larger flame. This shows that a simple lens can work when a formal tool isn't available, as long as you keep the tinder dry and the focus steady.

Understanding and applying benign exothermic reactions

Step 1. Know the idea. Some chemical changes release heat (exothermic) as they occur, while others absorb heat (endothermic). Rationale: recognizing heat release helps you separate reliable heat sources from those that won't help you start a fire.

Step 2. Keep safety first. Only observe safe, simple demonstrations in controlled, small-scale ways and avoid any unfamiliar mixtures. Rationale: heat makes flames, but mismanaged reactions can create dangerous fumes or pressure.

Step 3. Use only widely understood, non-harmful substances. Do not mix chemicals you don't fully understand or can't safely manage. Rationale: simple, well-known interactions reduce risk and keep you in control of the situation.

Step 4. Don't rely on chemistry as your main fire method. In the field, relying on safe, consistent methods that you can practice regularly is wiser than chasing a reaction that may not work when you need it. Rationale: practical firecraft emphasizes reliability and safety over flashy demonstrations.

Step 5. Observe and learn from natural heat. If you notice heat from natural processes, like metal warming in the sun or the slow warmth of a damp log drying in a sunny spot, reflect on how heat moves and how you might use it safely for ignition or drying tinder. Rationale: understanding heat flow improves judgment for real-world situations without creating risky experiments.

Two brief examples or micro-cases of this point

Example C. You hold a small metal scoop in the sun and notice it warms quickly. It isn't enough to ignite tinder on its own, but you observe that heat travels from metal to air. You use that idea to inform your judgment: concentrate heat with a lens on dry tinder rather than attempting a risky chemical reaction. This reinforces the principle that heat can come from many sources, but control and safety matter most.

Example D. In a safe, supervised setting, you compare rusting iron's slow heat with a lens's fast heat. The rust gives you a sense of how heat develops over time, while the lens demonstrates how to create a focused blaze quickly when conditions are right. You note that reliable ignition usually comes from plain, tested methods rather than experimental chemistry under field conditions.

Selecting scenarios where concentrated heat outperforms other methods

Step 1. Look for bright sun and dry tinder. A clear day with plenty of sunlight makes concentrated heat a strong option. Rationale: sun energy is abundant, accessible, and free; dry tinder catches with less effort when heat is concentrated.

Step 2. Ensure stable, safe surroundings. A steady stance, level ground, and an unobstructed clearing help you keep the lens focused and the fire bed undisturbed. Rationale: a stable environment reduces mistakes and increases your odds of success.

Step 3. Check the availability of a lens or improvised lens. If you have a magnifier or a water-filled bottle, you have a direct path to focal heat. Rationale: having the right tool in hand makes the technique reliable, not just possible.

Step 4. Read the conditions and the material. If the tinder is already damp or you're in a windy area, concentrated heat may not work well enough. In those cases, switch to other reliable methods you've practiced. Rationale: knowing when to switch keeps you from wasting time or getting frustrated.

Step 5. Use concentrated heat to initiate a small ember, then switch to a full fire-building method. Rationale: starting with a controlled ember helps you build a steady flame without overwhelming the tinder bed.

Two micro-cases showing these decisions in action

Example E. A bright day on an open field, you find dry grass and twigs. You focus sunlight with a clean magnifier onto the tinder, the patch sizzles, and soon you have a tiny ember that you feed into a protective fire bed. You've used the sun's energy efficiently and moved to the next step with a stable ember.

Example F. In a shaded campground with soft ground and damp tinder, you realize the sun is not strong enough to concentrate heat reliably. You shift to friction or sparks instead, applying the methods you've learned in other chapters. You still keep the lens handy for emergencies, but you don't force a method that won't work in the moment.

Key Takeaway: Concentrated heat using lenses or formed focal points works best with dry tinder, a steady setup, and safe handling. When the sun is strong and conditions are right, you can ignite an ember quickly; when it isn't, switch to a method you know well rather than forcing a heat trick.

Coming up next Subtittle: Firecraft in Adverse Weather and Terrain

1) Constructing windproof setups and micro-shelters,

2) Starting fires with damp, frozen, or waterlogged materials,

3) Managing smoke, moisture, and thermal efficiency in harsh conditions.

Chapter 6

Firecraft in Adverse Weather and Terrain

You will learn to start a fire in difficult weather by building windproof setups, lighting damp or frozen materials, and managing heat and smoke so you can cook, stay warm, and stay safe.

Windproof setups and micro-shelters

The first move in harsh weather is to reduce wind at the fire. Wind doesn't just blow flames away; it also carries heat away and makes smoke billow, which can make it harder to see and breathe. Start by finding a natural windbreak if you can—a rock outcrop, a line of trees, a low ridge, or even a snow drift can shield your flame. If you have none, build one using what's around you. Gather dry sticks, logs, or rocks and arrange them in a loose circle or wall to break the gusts. The goal is to interrupt the air flow enough so you can keep the flame stable without smothering it.

Next, create a small, simple hearth. A compact ring of stones or a shallow pit can keep your fire from spreading and protect the fuel from gusts.

Place the fire toward the leeward side of your windbreak, so the heat is directed toward your fuel pile rather than blown sideways. If you have a tarp, blanket, or a large leaf, you can fashion a tiny shelter behind the fire to trap warmed air. The shelter does not have to be fancy—a low lean-to or a curved wall of pine bines, branches, or a mound of earth works well. The idea is to limit direct wind while allowing enough airflow to feed the fire with oxygen.

A heat reflector increases efficiency. A simple reflector is a curved wall of dry stones or a low, smooth, dry surface placed a short distance behind the fire. The reflector catches heat and redirects more of it toward your kindling and fuel. If you're in snow, you can set the fire against a compact bank of snow to form a natural reflector. The snow also acts as a moisture barrier, keeping colder air away from the flame. Always test the wind direction and watch how it shifts. If gusts suddenly rise, you can adjust the wall, move the launcher area, or add more shelter to maintain stability.

Ground and surface matter too. Build on dry ground—bare soil, dry sand, or a cleared patch of earth—and keep a tidy radius around the fire. Remove flammable debris from within several steps and keep the fire small enough to control. Wind and moisture make the surrounding environment dangerous, so you should be extra mindful of embers, hot rocks, and hot coals that could flare up when the wind shifts.

Three quick practice steps help you get reliable windproofing fast: First, scout for a natural barrier or position. Second, assemble a small hearth with shields behind and around the flame. Third, place a reflector in the best position to push heat toward the fuel and cooking area. In steady weather, you can lighten the setup and simplify, but in storms or gusty days, don't skip the windbreak and the reflector—these are what keep your fire alive.

Two short micro-cases show how windproofing works in practice. Case A: You're on an exposed hilltop with a steady crosswind. You locate a small rock wall and build a knee-high ring to block the wind. You set the hearth in a shallow pit and position a curved log behind the flame to act as a reflector. The fire catches quickly, the smoke drifts away from you, and you can add small twigs with confidence. Case B: You're hiking along a riverbank where gusts funnel between trees. You use a fallen trunk and a pile of dry leaves to form a quick shelter, then prop a flat rock behind the flame and place a second, higher reflector to trap heat. The setup stays stable, and you can transfer heat to a pot without the wind snuffing the flame.

If wind shifts or rain starts, adapt fast. Push the fire closer to the shelter, lower the flame to reduce gust exposure, or add another layer of shelter. If the fire begins to smoke heavily, put the air into it by lifting a little of the barrier to let in more oxygen, then re-seat the reflector to guide heat back toward the fuel. If you do not have a shelter and wind becomes dangerous, stop and wait for calmer conditions. Fire safety comes first.

Troubleshooting windproofing: If the flame flickers and dies with gusts, your barrier may be too low or not properly oriented. Raise the barrier or shift it to block the wind direction more completely. If smoke chokes you, reposition the fire downwind of the windbreak and ensure there is space for the smoke to escape away from your face. If you are in wet conditions, dry materials are your friend. Use dry inner bark, dry needles, or dry wood that you find tucked in crevices or under larger logs. Do not force a flame with damp fuel; instead, conserve energy and switch to a drier spot if possible.

Starting fires with damp, frozen, or waterlogged materials

When the weather is unfriendly, dry tinder becomes the difference between a spark and a stubborn, stubborn glow. Your goal is to create a small, reliable nest that can catch a flame even if the world around you is wet. The process starts with finding or creating dry tinder and then building the smallest possible starter fire that can grow.

Begin with a simple, dry tinder nest. Fine fibrous materials like dry grasses, cattail fluff, pine needles, or shredded bark curls can light quickly when they're dry and dense enough to hold a spark or ember. If you have access to dry inner bark or dry bark fragments on a damp tree, those can also serve as tinder. The key is to keep it light, airy, and dry. You want the nest to hold a flame without smothering it, so keep the fibers loose enough to allow air to flow but compact enough to stay together once they begin to burn.

Next, locate or create dry kindling. Break larger damp pieces into smaller parts; sometimes splitting wood reveals drier material inside. If you can't find dry wood, look for wood that is less waterlogged, such as a branch with a dry core inside. Build a small starter structure using your tinder nest as the base and then place a few tiny sticks over it in a teepee or a V-shape so air can move through. The first flame will be fragile, but it should be enough to heat and dry the surrounding fuel, turning damp wood into usable heat.

A practical approach for stubborn damp conditions is to use the flame itself to dry the fuel. Once your tinder catches, feed it gradually with dry kindling to grow a stable flame. Do not rush to add large pieces; let the heat accumulate and dry the next layer of wood before you try to feed in bigger fuel. You can gently shelter the fire from wind to reduce heat loss and to concentrate the flames on the kindling. As the fire grows, it will create its own warmth and begin to dry the surrounding fuel, making it easier to add larger pieces.

Wood choice matters. Resin-rich woods can flare up and help in a pinch, but they can also burn with more smoke. If you're facing damp conditions, prioritize dry, hardwoods or wood with a dense, straight grain when you've got drier pieces available. If you must work with wet material, target the driest parts you can find and harvest them in small, quick stages so you don't waste energy.

Two micro-cases illustrate this method. Case A: You're in a soggy forest with damp sticks lying around. You split a damp log to expose the dry heartwood inside, then place a tinder nest and a tiny handful of dry twigs on top. A light breeze helps, and the ember from the tinder catches enough heat to dry the log into a flame. Case B: You're on a lakeshore where the air is cold and the wood is wet. You gather dry inner bark from a dead branch, shave tiny splinters from it, and construct a tight nest with the smallest sticks arranged in a cone. The tiny flame becomes a steady, small fire that can be expanded with more dry fuel once the weather warms or dries out.

If your tinder won't light, you can try a few ideas that don't rely on novelty but rely on solid technique. Make sure your tinder nest is dry and that you've shaved down the fibers to expose rough surfaces that grab the flame. Use a striker to deliver a focused spark if you're using a spark-based ignition, or carefully fan the fire with a hand or a piece of bark to inject oxygen, which can save a stubborn ember from fading away. If the flame still won't take, step back, re-examine your nest, and try a fresh portion of tinder or a new angle for the spark. You can also move the fire to a more sheltered micro-site to keep the heat contained in a smaller space.

In tough conditions, you'll sometimes have to be patient. Lighting a fire from damp materials often takes multiple tries, a calm mind, and a willingness to rearrange fuel and air. The ability to persist with a small, controlled flame will save you time and energy later when you need a fire for warmth, signaling, or cooking.

Two practical reminders help you stay on track. First, never force one method when moisture is high. If your attempt stalls, switch to a lighter, drier location, or try another tinder type you know dries quickly. Second, aim to create a small, stable flame first, then gradually add fuel. A big flame in damp air can burn up moisture faster but collapse as soon as fuel is too damp to ignite. Patience and careful staging win in wet weather.

Key lessons from this section reinforce the idea that preparation and choice of materials predict success. Look for dry tinder in pockets or exposed parts of the environment, and always be ready to adapt your plan when the weather shifts or your fuel quality changes. Even a small flame, carefully tended, can grow into a solid fire that keeps you warm and safe.

Managing smoke, moisture, and thermal efficiency in harsh conditions

In very cold or humid environments, smoke can be a frequent companion. Smoke is more than an annoyance; it hides air flow problems and can signal that fuel and oxygen aren't meeting in the right way. The remedy is to optimize airflow, locate a comfortable position relative to the wind, and maintain a steady rhythm of feeding the fire with progressively larger fuel. When smoke rises heavily, adjust the fire's angle and the barrier around it so the smoke can escape away from your face while you still get enough heat.

Moisture management is the other side of the coin. Moisture coming off the fuel lowers heat, so you want to keep a dry zone around the fire and use a reflector or windbreak to shield the flame from rain or mist. If you can, place your fire on dry ground or build a small dry stand using logs or flat rocks to raise the fire off the wet surface. A dry base reduces steam and heat loss, making the flame easier to start and sustain. Another simple trick is to cover the fire lightly with a layer of dry leaves or pine needles to trap heat while still allowing air to circulate, which helps the fire maintain a steady burn in damp situations.

Thermal efficiency comes from balancing input and recovery. The fire needs oxygen to burn cleanly, but too much wind drains heat; too little oxygen leads to black smoke and slow burning. The solution is to lean toward a small, well-ventilated flame rather than a roaring blaze in a rainstorm.

The smaller flame is easier to protect from the wind and easier to feed with dry material as it dries. A reflector can push heat toward the fuel, and an improvised windbreak can help maintain a steady burn. When done correctly, you'll notice less smoke, less fuel wasted, and more heat staying close to your pot or your hands.

A few common pitfalls and fixes: If the fire smokes a lot, check air supply and remove damp fuel; if possible, position fuel away from the wind to allow oxygen to reach the flame. If heat is escaping upward and away from your cooking area, reposition the reflector and the fire so heat moves toward you rather than up and over. If moisture accumulates on the fuel and you see steam, pause and let the fire dry the material a little before continuing. If embers go out, re-create a small tinder nest and reintroduce tiny, dry fuel slowly to restart.

Examples and micro-cases to illustrate the fourth dimension of weather resilience: Case A shows a snowy hillside where moisture and wind threaten the flame. You arrange a low wall of dry sticks and stones behind the fire, nest tinder in a dry cavity, and use a curved piece of bark as a heat reflector. The flame stabilizes; smoke becomes a helper to monitor wind changes. Case B places a fire in a damp forest clearing. You choose dry inner bark, build a tight tinder nest, and shelter the fire with a simple rock and leaf canopy. The small flame grows into a steady blaze, and you maintain heat without wasting fuel.

In all cases, the goal is steady heat while keeping people safe and comfortable. The weather can be stubborn, but with windproofing, careful material selection, and mindful airflow, you can turn a difficult environment into a reliable place to light a fire, cook a meal, and stay warm.

Key Takeaway: In harsh weather, windproofing and sheltering your fire are the first moves; they protect the flame, conserve fuel, and set up every other step for success.

Coming up next Subtittle: Urban and Improvised Ignition Sources

1) Repurposing everyday items as starters, reflectors, and windbreaks,

2) Safely extracting ignition potential from batteries, wires, and metal,

3) Reducing smoke and managing visibility for safety and comfort

Chapter 7

Urban and Improvised Ignition Sources

You will learn three practical ways to use everyday items and common materials to start fires safely in urban or improvised settings, while keeping smoke low and visibility controlled.

Repurposing everyday items as starters, reflectors, and windbreaks

In urban or improvised settings, you can turn common objects into tools that help you light a fire without needing modern gear. The idea is to use what's already around you, keeping safety and consideration for others at the center. Think of simple metal items, glass, cardboard, and sturdy plastics that can be repurposed to help you start and manage a small, controlled flame.

1) Find a reflective surface and line up the sun

Look for clean, smooth metal surfaces—like a small lid, a metal tray, or even a tight sheet of aluminum foil. The goal is to reflect sunlight onto your tinder bundle, concentrating light and heat to nudge it toward ignition. Adjust the angle slowly so the bright spot sits where your tinder rests, not on your hands or your clothing. The more focused the heat, the better your chances of a spark forming without scorching your fingers. The rationale is simple: sunlight is a ready-made heat source, and a reflective surface makes that heat more efficient.

2) Shape a compact reflector or solar funnel

If you have a sturdy lid or a small metal cup, bend or angle it to direct sunlight onto a small patch of tinder. A smooth, curved surface can create a tiny solar focal point that warms the tinder gently but steadily. If you can't bend metal, use a clean piece of glass or a reflective sheet to achieve a similar effect. The principle here is that light can be concentrated with a well-placed surface, giving you a sparing but real chance to ignite dry material.

3) Use windbreaks to steady heat and protect the tinder

Wind can steal heat and chill the tinder, so build a simple windbreak from bricks, rocks, or a few sturdy sticks arranged to form a little wall. The windbreak reduces cooling, keeps smoke away from you, and makes ignition more predictable. Position the reflector so it faces the tinder with the wind coming from the side or behind you, never directly at your face. The rationale is safety plus a steadier burn; a calm pocket lets the tinder glow without being blown out or extinguished by gusts.

4) Dry, ready tinder and a stable base

No matter what you repurpose, your tinder must be dry and fluffy enough to catch a spark or a focused heat. Keep it in a small natural container or a dry pocket until you're ready. Place your tinder on a non-flammable base—rock, dirt, or a metal tray—to avoid accidental scorching of surfaces. The principle is preparation: dry tinder and a stable stage increase your odds of a successful ignition with modest heat.

Examples in action

Example 1: In a sunlit courtyard, you lift a clean metal lid and angle it to reflect a bright spot onto a small nest of dry tinder made from dry grass and bark shavings. After a few minutes of steady warmth, the tinder glows and a tiny flame catches, then you feed it with tender strips of wood laid in a gentle teepee shape. The moment you see flame, you control the burn by adjusting airflow with your windbreak and by gradually adding larger pieces.

Example 2: On a shaded balcony, you use a piece of polished aluminum foil taped to a shallow tray to create a little solar reflector. You place a tiny pile of dry pine needles beneath it and orient the foil toward the sun. The needles heat and catch a spark from your tinder bundle, giving you a stable flame that you then draw into a controlled bed of coals with careful feeding and minimal smoke.

Safely extracting ignition potential from batteries, wires, and metal

Urban and improvised settings can include items that store or conduct energy in small ways. You can study the basic idea—how certain metals and connections can produce heat or sparks—without rushing into unsafe experiments. The focus is safety, restraint, and awareness: never press metal against skin, never short-circuit damaged devices, and always keep a clear space around you and others.

1) Inspect items for damage and decide if experimentation is appropriate

Check batteries, wires, and metal parts for cracks, leaks, bulges, or corrosion. Damaged batteries can vent dangerous chemicals or heat up quickly. If you see damage, do not attempt to use the item for ignition. The underlying principle is caution: avoid handling components that could fail safely, and choose safer alternatives when possible.

2) Separate and protect metal contacts

If you're exploring the ignition potential in a classroom or outdoors with supervision, keep metal contacts separated when not in use. Use insulating barriers or gloves to reduce the risk of skin contact in case a spark or heat occurs. The idea is to create a safe boundary that prevents accidental burns or short-circuits while you observe how materials respond to energy.

3) Observe, don't force ignition

You may study how stored energy can behave in theory by listening for the telltale crackle of a tiny spark or the gentle warming of a metal surface, but avoid attempting to produce flame from damaged or unfamiliar devices. The purpose is learning, not improvising fire from risky sources. If you must test an item, do so in a controlled, safe setting with appropriate supervision and equipment that protects you and others nearby.

4) Use safer energy demonstrations when possible

Whenever a demonstration is possible without ignition, use it. For instance, you can observe heat buildup on a metal strip by rubbing it slightly or by placing it near a heat source and watching how the surface changes. The rationale is to understand energy behavior without creating a hazard, so you can apply the idea more safely in real firecraft later.

Examples in action

Example 1: In a supervised workshop, you examine a small, intact battery with a protected connector and a short length of insulated copper wire. You carefully observe how linking the two ends in a test cradle can generate warmth at the contact point, but you stop well before any direct contact that could spark or heat a material to ignition. The exercise builds intuition without risk.

Example 2: In a park, you lay out a metal strip with insulation and demonstrate how heat transfer can occur from a safe heat source to a surface. You note where heat concentrates and how a small patch of tinder placed nearby would respond if ignition were attempted, always emphasizing safety and the non-endorsement of unsupervised experimentation with real flames.

Reducing smoke and managing visibility for safety and comfort

Smoke can cause breathing irritation and reduce safety by signaling unwanted attention. Keeping smoke low helps you stay aware of your surroundings and makes fires more controllable in urban spaces. You can manage smoke by choosing appropriate fuels, controlling airflow, and shaping the fire structure to burn cleanly and steadily.

1) Choose dry, less resinous fuels and keep them organized

Dry wood, small twigs, and tinder that burns cleanly produce less smoke than damp logs or resinous materials. Keep fuel arranged so you can add pieces gradually, avoiding big, heavy loads that smolder and emit more smoke. The principle is simple: cleaner fuel and controlled feeding equals less smoke and easier visibility management.

2) Create a compact fire structure to limit draft

Build a small, efficient fire frame—think a tight teepee or a compact lean-to—that allows you to control airflow. Too much airflow can pull smoke toward you; too little can smother the flame. The goal is a balanced burn where heat rises efficiently and smoke remains behind or to the sides, away from your line of sight. The rationale is to maintain a safe, visible fire while keeping the air as clear as possible.

3) Use windbreaks and position for safety

Windbreaks protect your flame from gusts that would scatter smoke and blow embers. Place windbreaks on the side from which wind usually comes, not directly in front of you where smoke concentrates. The underlying idea is to create a sheltered pocket where the fire can burn with consistent warmth and minimal smoke bloom.

4) Manage ignition size and flame progression

Start small, add fuel slowly, and monitor how the fire behaves. If you notice heavy smoke, reduce the amount of material burning at once and open a little more airflow only if it helps the flame catch cleanly. The principle is conservatism: a smaller, cleaner burn is easier to control and lasts longer with less smoke.

Examples in action

Example 1: On a breezy rooftop at dusk, you place a windbreak behind a small teepee of tinder and kindling. You select dry pieces and feed them gradually, keeping the flame contained and smoke focused away from the building's edge. The result is a stable glow that doesn't draw too much attention while still producing enough heat for a gentle boil or warmth.

Example 2: In a backyard camping setup, you use a metal lid as a reflector to direct the heat toward the base of your tinder, reducing the amount of smoke that rises into the air. With careful feeding and a small, steady flame, you maintain visibility and control while keeping neighbors and family comfortable.

Troubleshooting quick fixes

If smoke is heavy and hard to control: check fuel dryness, reduce the amount of material burning at once, and adjust windbreaks or airflow to balance the flame. If ignition is slow or stubborn: ensure tinder is dry, refine the fire structure, and shield the flame from gusts while gradually feeding small pieces. If you're unsure about the safety of an item:

stop, reassess, and choose a safer alternative like sun reflection or dry tinder rather than testing a risky material.

Templates and tools

Urban ignition safety checklist:

1) Inspect all items for damage; discard anything cracked or leaking.

2) Confirm tinder is dry and ready for ignition.

3) Build a small, stable windbreak and position a reflector properly.

4) Check wind direction and set your fire structure to balance heat and smoke.

5) Keep a water source or sand nearby and know how to put out the fire quickly if needed.

Key Takeaway

Simple, safe improvisations with everyday items can help you start small fires with less smoke when you stay focused on safety, dryness, and controlled airflow.

Coming up next: Building Efficient Fire Architectures

1) Choosing between teepee, lean-to, log-cabin, star, and reflector builds,

2) Feeding strategies to control intensity for cooking, warmth, or signaling,

3) Transitioning from spark to sustained bed-of-coals.

Chapter 8

Building Efficient Fire Architectures

You will be able to choose and set up efficient fire architectures for cooking, warmth, or signaling, control heat with practical feeding methods, and reliably move from a spark to a sturdy bed of coals.

1) Building efficient fire architectures

When you start, pick an architecture not just because it looks neat, but because it fits the wind, the fuel you have, and what you want the fire to do. Each shape has its own rhythm and purpose, and knowing when to use which helps you get flame quickly, keep heat steady, and protect the flame from gusts. The common shapes are teepee, lean-to, log-cabin, star, and reflector, and each one changes how air moves, how heat concentrates, and how you can feed it without smothering the flame.

Teepee builds rise quickly from a tight center of tinder and kindling into tall flames that draft well. They light fast when tinder catches, and they're forgiving for beginners because the flames draw air from all sides. The downside is that a teepee burns hot and fast, which is great for ignition but can burn through small tinder and kindling quickly if you don't manage it. If you plan to cook later, you may need to shift to a more controlled structure once the bed of embers forms.

Lean-to builds lean a single direction, usually against a natural shielding surface like a rock or a log. This shape protects the flame from wind and focuses heat toward the cooking area or the spot where you want warmth. It's steadier in gusty weather than a teepee and makes it easier to keep a small, steady flame as you add fuel. The trade-off is that air has to travel through the lean-to in a more linear way, which can slow ignition if you don't set up the initial fuel stack correctly.

Log-cabin arrangements stake two or more sets of fuel as a square or rectangular box. Logs are stacked in pairs, creating a stable center that holds heat longer. This shape is excellent when you want a longer burn for cooking or keeping a chill away for a while. It takes more time to ignite than a teepee, but once the coals form, you have consistent heat with less tending. The key is to place the tinder and kindling in a way that the inner core can glow without the outer logs starving it of oxygen.

Star patterns radiate lines of fuel outward from the center like spokes on a wheel. The flame sits in the middle and the fire grows outward along the spokes as air reaches deeper layers. A star can be useful in very windy spots because the outward branches shield the center. It can be trickier to balance to keep the center hot enough for ignition, and you may need to feed more deliberately to keep the core from cooling too much.

Reflector setups use a flat surface—like a clay wall, a rock, or a metal plate—placed behind the fire to bounce heat toward a cook or a shelter area. This can dramatically increase the perceived heat on the target zone without visibly increasing flame height. Reflectors are great when you have a thin flame or you're trying to warm a pot from behind the heat source. They're not ideal if you need a large, fast-burning flame for ignition or if you lack a stable, reflective surface.

How you choose comes down to conditions. If the wind is light and you want speed, start with a teepee to build a quick, bright flame. If wind is persistent, lean-to or reflector can protect the flame while you heat a pot or warm hands. For longer cooking or night warmth, a log-cabin gives you steadier heat with less frequent tending. In a windy, open space, a star pattern can distribute heat while defending the center, and a reflector behind any of these shapes can magnify the heat toward the cooking or camping area.

Example 1: A calm morning on a lakeside. You set up a teepee with small sticks and a generous bed of dry tinder in the center. The flames rise quickly, igniting the kindling. After a few minutes, you see a steady core of embers forming. You transition to a small log-cabin layout by moving the logs around the center as embers appear, producing a reliable heat source for warming a kettle. The sequence from ignition to steady heat happens smoothly because you started with rapid flame and then moved to longer burn as heat concentrated in the core.

Example 2: A windy hillside in the late afternoon. You choose a lean-to against a rock face, with a reflector to protect the space behind the fire. The wind chases away some heat, but the lean-to keeps the flames directed where you need them. The reflector jumps extra heat toward the cooking area, and you can keep a pot simmering with minimal flame height. The fire remains steady, and you avoid smoky gusts that could blow into your eyes.

Two quick notes on safety as you shape and feed your fire. Always clear a safe perimeter around your fire site. Remove dry leaves, pine needles, and other easily flammable debris within at least a couple of meters. Have a bucket of water, sand, or dirt nearby, and don't leave the fire unattended. When you're done, let the embers fully cool before you move or stash the fire away. If you're in a protected area or a place with rules, respect local guidelines about fire use.

2) Feeding strategies to control intensity for cooking, warmth, or signaling

Heat and flame aren't the same thing. You'll want to adjust the size, the pace, and the placement of fuel to match your goal, whether it's cooking, staying warm, or signaling others from a distance. The guiding principle is to feed gradually, watch how the flame responds, and adjust before the fire runs away with you.

First, define your goal. For cooking, you want steady, contained heat with small, controllable flames that won't scorch your food or burn your pot. For warmth, you want a longer burn with a steady bed of embers you can bank and maintain without constant tending. For signaling, you want bright, visible flame and heat that can endure the conditions so someone can spot you from a distance.

Next, choose fuel size and shape. Fine tinder catches quickly and starts the flame; small kindling keeps the flame alive and grows the fire to embers; larger fuels sustain heat without needing to crank the flames higher every few minutes. A general rule: start with tinder and small kindling, add mid-size sticks to grow the flame, and finally introduce larger logs only after you've established a healthy bed of embers that can glow for a while.

Finally, practice smart feeding. Feed in small amounts rather than large clumps. Place fuel into the fire so it can catch air and burn efficiently; avoid stuffing too much material into a flame at once, which can suffocate it. For cooking, a common approach is to build a backbone of small logs arranged to let air move through and around them, then cap the area with slightly larger pieces to shield direct flames from delicate pots. For warmth, you'll keep a steady chain of tiny additions on the outer edge so the core can glow while the fire stays controllable. For signaling, you may want short bursts of brightness followed by a calm glow to maintain visibility for onlookers who can be several hundred meters away.

Two brief examples show how feeding strategies play out in real moments. Example A: You intend to heat water for emergency tea. You start with a teepee and keep feeding small sticks as the flames rise. Once the surface shows bright, contained heat and a ring of embers forms, you add a slightly larger log on top, just enough to let the embers keep glowing under the pot. The flame height remains manageable, and the water boils without a freestanding flame licking the sides of the pot. Example B: You want a warm camp for the night and a signal to others if needed. You craft a log-cabin arrangement to sustain heat for hours, feeding regularly with small amounts of wood to keep the bed of coals alive. If visibility is important, you add a few longer branches to the edge to create a visible glow that catches eyes from across a clearing.

If wind or moisture complicate feeding, slow your approach. A little extra time spent arranging the fuel, testing airflow, and watching the way the flames respond yields better results than rushing to load in big logs. In damp environments, you may need to pre-dry some small pieces and rely on a reflector to push more heat toward the cooking area. In rain or snow, shield the fire with a natural windbreak and maintain the flame with frequent but small feedings to preserve oxygen flow. The goal is steady heat that you can control with a calm hand and a patient eye.

3) Transitioning from spark to sustained bed-of-coals

From the moment you light the tinder, you're steering toward a bed of glowing coals. The transition is the critical moment when the fire shifts from bright flames to a stable, heat-rich core that can support cooking, warmth, or signaling for longer periods. You'll move through a small sequence, and with practice, you'll be able to judge when the bed is ready for its next step.

Step 1: Light and secure a flame. Light the tinder and allow the flame to catch the smallest kindling. Keep the flame focused, not wobbly, and ensure there is enough air movement to feed the initial growth. If the wind threatens, adjust your shelter or relocate to a wind-protected area. A secure flame is the foundation for a strong transition.

Step 2: Grow the ember base. After ignition, place a few layers of progressively larger material around the core so air can circulate and the inner layer begins to glow. The goal is a steady glow rather than a roaring flame. If flames rise too high or if the flame keeps snapping upward, reduce the size of the fuel pieces you add and shift to a more compact arrangement that channels heat into a bed of coals.

Step 3: Form the bed of coals. When the center glows with a steady charcoal color and there is a steady, low flame around the edges, you've formed a bed of embers. At this stage you can cook directly on the coals or place a reflector plate to shine heat toward your cooking surface. If you see bright yellow flames dominating, ease off feeding for a moment and allow the glow to catch up. Handle this transition with patience; forcing it too quickly can flare the flames out of control.

Two short micro-cases help illustrate success in this transition. Micro-case A shows a beginner moving from a small teepee to a reliable bed of coals for a cooking pot. The youngster lights tinder, adds a little kindling, keeps a steady heat, and when a ring of glowing coals forms, places a small pot directly over the bed. The result is a gentle, predictable heat that doesn't scorch the food. Micro-case B demonstrates a cold evening where a reflector-backed log-cabin structure is used. The initial flames rise with a stable core, and as the coals build, the fire sustains warmth and a strong glow that can be banked for hours if necessary.

Troubleshooting notes

- If the flame goes out: Check for dry tinder, ensure there is enough oxygen by not crowding the center, and re-light with fresh tinder if needed. If there is wind, shield the flame with a natural barrier or reposition.

- If you get soot or smoke: Improve airflow by rearranging the fuel so air can reach the heart of the flame, or reduce the amount of moisture in your wood by drying it or substituting drier pieces for a moment.

- If heat is uneven or the bed doesn't form: Adjust the structure to bring the center into a better burn path; avoid crowding fuel and allow space for air to flow into the center as you add more fuel.

Tools and templates you can copy

- Fire architecture quick-choice guide: If wind is light, consider teepee or log-cabin for heat; if wind is strong, lean-to or reflector behind the fire helps maintain the flame; if you need concentrated heat for cooking, start with teepee then shift to log-cabin with a reflector and a careful feed.

- Simple feed plan: Start with tinder and small kindling, add mid-size sticks as the flame stabilizes, then bring in larger logs gradually. Keep a calm cadence and monitor heat. For signaling, keep sections of brighter flame visible at intervals, then balance with a calmer glow.

- Lighting and transition checklist: 1) Clear perimeter and prepare tinder. 2) Build the initial structure. 3) Light tinder and gently feed. 4) Watch for embers developing. 5) When the bed glows, shift to cooking or warmth mode as needed.

Key Takeaway: A well-chosen fire architecture, paired with disciplined feeding and a careful transition to a bed of coals, gives you reliable heat for cooking, warmth, or signaling without unnecessary flame.

Coming up next Subtittle: Ember Management and Long-Term Heat Strategies

1) Creating, transporting, and protecting live embers,

2) Reigniting from smoldering coals and conserving fuel,

3) Using thermal mass, bed cooking, and slow-heat techniques.

Chapter 9

Ember Management and Long-Term Heat Strategies

By the end of this chapter, you will be able to create, transport, and protect live embers, reignite from smoldering coals, and use thermal mass and slow-heat techniques to keep heat steady and useful.

Creating, transporting, and protecting live embers

You start with a clear goal: keep a small, glowing ember alive long enough to grow into a sustainable fire. Take a moment to choose a safe, dry spot for your ember work. Look for a sheltered nook in your fire area, away from dripping branches, damp ground, and places where wind can gust directly through your tinder bed. Wood ash is helpful here because it slows cooling and can shield embers from sudden drafts, so creating a thin ash ring around the ember zone is a smart move when feasible. You'll want a very small, steady glow rather than a roaring flame, because a steady glow preserves energy for later.

Step 1: Create a tiny ember core. Place a compact nest of dry tinder in the center of your prepared spot. The tinder should be dry and fibrous—think fine, easily caught bits like dry bark, twigs, or needles. A compact glow is easier to protect than a big flame. Rationale: embers need steady heat, not continuous high flame, to survive long enough to be used again.

Step 2: Nurture with careful fuel. Add only small amounts of dry fuel at a time—thin sticks or slivers—just enough to keep the ember alive without smothering it with flame. Rationale: too much fuel at once cools the ember or chokes it with smoke, but too little won't sustain it.

Step 3: Shelter the ember from wind. If wind is a factor, create a simple windbreak with a flat rock, a small piece of bark, or a natural barrier such as a low wall of dirt. The goal is to slow the blast of air that can blow your ember out. Rationale: wind makes embers flicker out or burn hot too fast, but a gentle shield keeps the glow intact.

Step 4: Transport ember with care. If you need to move the ember to the main fire or to a new location, use a metal tool or gloved hands to lift the ember bed—never drag it with bare skin over hot surfaces. Place it gently into a prepared bed or onto a layer of dry tinder. Rationale: rough movement or direct contact with hot coals can break the ember and end its life.

Step 5: Protect during transfer. If you must carry an ember across gaps or over uneven ground, keep it covered with a small, dry container or shield it with a dry leaf to guard it from gusts and moisture. Rationale: exposure to rain or damp air ends the ember's life quickly.

Step 6: Keep a pocket ember supply. Have a small stash of dry tinder and a few short sticks ready so you can quickly feed and grow the ember when needed. Rationale: quick access to dry fuel makes the transition from ember to full fire smoother and speeds up the process.

Examples and micro-cases

Micro-case A: You're camping in a pine forest. A light breeze slides through your campsite. You clear a patch, lay a flat stone as a frame, and set a tiny nest of dry pine needles in the center. A soft glow appears. You steadily feed the glow with three tiny slivers of dry twigs, keeping the glow bright but contained. The ember sits sheltered by your stone frame for several minutes, maintaining warmth while you prepare a larger fire ring around it. When it's ready, you gently move the glowing core to the center of your larger tinder bed and begin building a proper flame with larger sticks.

Micro-case B: On a windy beach, you need to protect an ember while it grows. You improvise a windbreak from a flat piece of driftwood and a shell, shielding the ember nest beneath. The ember glows steadily for longer than expected. You transfer the ember to a prepared bed under the windbreak, feed it with a few dry shavings, and soon a steady flame leaps up without a rush of air blowing it out.

Reigniting from smoldering coals and conserving fuel

Reigniting from smoldering coals is about reading the fire's hidden signals and using air and tinder wisely to bring back bright flames without wasting fuel. The idea is to coax life from the coals rather than blasting them with energy and losing heat in the process.

Step 1: Read the coal bed. Look for spots that still feel warm to the back of your hand, not just the surface glow. A few warm spots indicate there's still energy inside the bed. Rationale: reviving a fire from these pockets saves fuel and time.

Step 2: Clear some ash to invite air. Gently move aside a thin layer of ash from over the warm spots to uncover them enough for oxygen to reach the embers. Rationale: air is what lights coals; exposing warm coal pockets helps flames reappear without needing a lot of new fuel.

Step 3: Add small, dry tinder. Lightly place a small amount of dry tinder (fine shavings, dry bark, or pine needles) over the warm pockets. Rationale: tinder catches quickly and creates a small flame that can spread to larger fuel.

Step 4: Introduce controlled airflow. Use a soft puff of air with your mouth or a hand fan to feed the tinder flame, then remove the source of disturbance as soon as the flame takes hold. Rationale: a gentle, controlled supply of oxygen is more effective than a blast of air that can scatter embers and waste heat.

Step 5: Build gradually, don't flood it. Once the flame appears, add very small pieces of dry wood in stages. Let each addition burn a moment before adding more. Rationale: gradual feeding keeps the heat stable and prevents smothering.

Step 6: Conserve fuel by reusing existing heat. If you have a bed of embers from the earlier session, you can reincorporate them by feeding them into the new flame rather than trying to start from scratch. Rationale: reusing heat is efficient and reduces the amount of new fuel you need.

Examples and micro-cases

Micro-case C: After a midday hike, you return to a fire pit with mostly smoldering coals. You inspect the bed, uncover a warm pocket, and breathe a careful puff across it. A thin yellow flame grows, then you place a scrap of dry bark over the flame and slowly add a few dry sticks. The fire re-establishes with minimal new fuel, and you're ready to cook.

Micro-case D: On a damp day, smoldering coal lies under a bed of damp ash. You carefully clear an area to expose warm coal, then place a small tuft of dry pine needles on it. The needles catch, the flame grows, and you quickly feed in a few more sticks, maintaining a steady burn rather than a roaring flame.

Using thermal mass, bed cooking, and slow-heat techniques

Thermal mass uses materials that store heat—like stones, dense earth, or compacted soil—to radiate warmth long after the flame dies down. Bed cooking uses a bed of embers to transfer heat to food or water with minimal direct flame. Slow-heat strategies keep heat moderate so you can cook, stay warm, or signal without wasting fuel.

Step 1: Build a heat-retaining bed. Gather flat stones or a shallow pit lined with sand or earth. Create a compact bed of embers in the center, with a ring of stones around it to hold heat. Rationale: a dense, well-placed bed will radiate heat for longer, letting you cook or stay warm with less fuel.

Step 2: Prepare a simple reflector and windbreak. Place a flat stone or a curved piece of metal on the windward side to reflect heat toward your cooking area. A low windbreak helps maintain a steady flame or heat line. Rationale: reflectors direct heat where you want it and reduce heat loss to the wind.

Step 3: Add a cooking surface that won't crack. Use a clean, dry rock or a metal surface if available to rest pots, pans, or even a metal can. If you're using stones, choose ones that won't pop or crack under heat. Rationale: damaged cooking surfaces can ruin your meal and create dangerous fragments.

Step 4: Move from flame to bed to conserve heat. When you're not actively cooking or warming, shift your setup from direct flame to the ember bed so heat continues to radiate without burning fuel at the surface. Rationale: using thermal mass and bed heat maintains warmth with less fuel.

Step 5: Manage fuel for slow-heat cooking. Push larger logs to the edges of the ember bed and keep smaller fuel close to the heat source. Maintain a steady, gentle burn rather than large flames. Rationale: slow, even heat is safer and more efficient for long cooking or keeping warmth.

Step 6: Safety first with heat storage. Avoid placing flammables near the hot bed and always monitor for sparks. If you notice any risk, reduce heat or extinguish and restart with more protective measures. Rationale: safety protects you and your surroundings.

Examples and micro-cases

Micro-case E: You want to cook a pot of soup with minimal fuel. You lay a circle of flat stones, build a small ember bed in the middle, and place a metal pot on the stones. A reflector stone funnels heat toward the pot. You maintain a gentle glow for forty minutes, adding a few small sticks only when the flame begins to taper. The soup simmers slowly, using the stored heat from the bed for stable warmth without constant feeding.

Micro-case F: On a cool evening, you collect a pile of dry pine needles and small sticks and create a compact ember bed beneath a curved board acting as a wind shield. You place a kettle on the makeshift stove and let the heat radiate from the bed through the grate. After a while, you reduce the fuel and rely on the bed's heat to finish heating your water. The result is steady heat with lower fuel use and less attention required.

Key Takeaway: The best fires start with a smart plan for energy. Protect a tiny ember, coax it to life, ride the glow with careful feeding, and let heat stay in place by using thermal mass and a stable bed. This approach saves fuel and extends your options for cooking, warmth, or signaling.

Coming up next: Ethics, Regulations, and Emergency Planning

1) When to avoid fire: environmental and legal considerations,

2) Leave-no-trace principles and minimizing ecological impact,

3) Emergency signaling, hypothermia prevention, and contingency plans.

Chapter 10

Ethics, Regulations, and Emergency Planning

Outcome: You will know when to avoid lighting fires, how to minimize ecological impact, and how to prepare for emergencies.

When to avoid fire — environmental and legal considerations

In many places, fires are not allowed at certain times or in certain spaces. You should treat every outdoor area as if rules exist even if you can't see a sign. The most common reasons to avoid fire are high fire danger, strong winds, dry or windy weather, and locations where fires are prohibited by law or by park rules. If you heat wood and later see smoke or embers traveling, that is often a sign that conditions are risky or the area is protected. Environmental rules exist to prevent wildfires that could threaten forests, animals, homes, and people. If you're in a national forest, state park, or nature reserve, you may find explicit bans posted at entrances, trailheads, or campground kiosks. Even when fires are allowed, there are limits: some places let only certain types of fires, such as contained stoves or metal fire pits, and there may be fuel restrictions or specific times when fires are permitted.

A practical approach is to check first. Look for local burn bans or fire restrictions on official government or park websites, or ask a ranger if you're in doubt. If you're traveling, call ahead or check signage at the trailhead. If weather has been dry for days or weeks, assume that a fire could be prohibited even if a sign looks permissive. If you see dead trees, compromised soils, or a landscape that looks unusually dry or fragile, those are signs to pause and reassess. If you're in a place with sensitive wildlife or fragile habitats, even a small flame can cause lasting damage. In short, if you aren't sure, you shouldn't light a fire.

Legal considerations go beyond simple bans. Some areas require permits for fires, some require reporting if you start one, and some areas regulate the size of the flame or the fuel you may burn. In many busy places, there are penalties for starting or failing to extinguish a fire properly. You may also have to consider neighbors and communities who could be affected by smoke, sparks, or risk of spreading fire to nearby structures. The safe rule is to treat any location with potential restrictions as if a rule might exist, and to verify before you act. If a location is off-limits, switch to a no-fire plan and focus on windbreaks, insulation, and emergency preparedness instead. If a fire is allowed but the conditions feel marginal, choose alternative methods or postpone the activity. Your goal is to keep people and the landscape safe while preserving your own ability to use outdoor spaces in the future.

Two micro-cases help illustrate how this works. Case A: You're camping in a well-marked state park during a dry spell. A ranger truck pulls up and reminds everyone that there is a limited burn window with strict restrictions. You pause, confirm the rules with the ranger or the posted signage, and decide to postpone any flame. You shift to using a stove or another non-flame cooking method, knowing you can still feed yourself while respecting the rules. Case B: You hike into a popular forest that has a daily fire ban posted at the trailhead. You arrive after a long day and are tempted to make a small campfire for warmth and cooking. The sign clearly states no fires, and you choose a no-fire plan, using a heat-retentive layer, emergency blanket, and a windscreen improvised from existing gear. The choice keeps you out of trouble and protects the area for others who come after you.

Quick checks you can run before you light anything

- Is there a fire ban in effect? If so, do not light a fire.

- What is the weather forecast? Is it windy or very dry?

- Am I in a protected area or a location with specific rules about fuels and pits?

- Do I know where nearby water sources and escape routes are in case conditions change?

- Do I have a plan to extinguish completely and leave no trace?

Leave-no-trace principles and minimizing ecological impact

Leave-no-trace is a standard mindset for outdoor ethics. It means making as little mark as possible and leaving the area looking as it did when you arrived. For fire, this translates into choosing a location wisely, using only resources you can responsibly manage, and ensuring you do not harm soil, vegetation, or wildlife. The simplest way to start is with respect for the land and a plan to minimize impact before you light anything.

First, use established fire rings or pits when they are allowed. If there is no fire ring, you might choose a stove or another safe alternative. If fire is allowed in a designated pit, do not create new pits or move rocks to build a new ring; stay within the established footprint. Gather wood nearby only in small amounts and avoid removing wood from living trees or delicate shrubs. Avoid piling fuel where it can scorch soil or burn uncontained. Keep the flame small, steady, and controlled, and never leave a fire burning unattended.

Extinguishing a fire is one of the most important parts of Leave-No-Trace. The rule of thumb is to treat a fire as if it will go out on its own and then check that it is completely out. Douse with water, stir the ash and coals, and douse again until there is no steam or warmth. If you don't have water, use dirt or sand until the area feels cold and you can touch it with the back of your hand without feeling heat. Do not leave glowing coals, and do not bury fire remnants in soil where they might re-ignite. Once the area is cool, scatter or pack out the ashes if required by local rules and pack out any litter, wrappers, or non-organic debris you brought in. Even if you are in a rush, take the extra minutes to clean up and leave nothing behind.

Two practical micro-cases illustrate Leave-No-Trace choices. Case A: You arrive at a forest clearing that has a small, established metal fire ring. You place your fire in the ring, keep the flame small, and avoid burning materials that release thick smoke. When you're finished, you drown the fire with water, stir, and drown again, then check that the area is cool before leaving. Case B: You hike into a sandy seam of ground near a riverbank with no ring and a sign prohibiting fires. You decide to cook using a compact stove built to burn cleanly and minimize smoke, and you pack all scraps and unused fuel away, leaving no trace of your visit.

Templates you can copy for quick use

- Leave-No-Trace Fire Check: Confirm burn ban status, inspect area for signs of previous fires, use an established ring if allowed, keep the flame small, gather only necessary fuel, extinguish completely, and pack out all litter.

- Fire Extinguishing Checklist: Douse with water, stir, douse again, feel for heat with the back of your hand, repeat until completely cold, and pack out ashes if required.

Emergency signaling, hypothermia prevention, and contingency plans

Emergencies can happen anywhere. Being prepared makes a big difference in outcomes. The core idea is to reduce panic, increase your visibility to helpers, and keep your body warm and safe while you wait for assistance or contact help yourself.

Signaling is about making yourself noticeable when you are lost or hurt. Simple, reliable signals include a loud, distinct noise such as three short blasts or three whistle blasts, which is a widely recognized distress signal. Use bright colors, a reflective surface, or a mirror to create flashes that can catch a rescuer's eye. If you have a light at night, use it as a beacon. Avoid placing yourself in risky situations just to attract attention. Instead, choose a safe, open area where you can be seen from a distance and stay put if you're unsure of your location, unless moving to a safer spot is clearly necessary.

Hypothermia is a real danger in many outdoor settings, particularly when temperatures drop, wind picks up, or shelter is poor. Prevention is easier than treatment and includes layering clothing, staying dry, and seeking shelter in a windbreak or shelter if possible. Signs of hypothermia can include shivering that won't stop, confusion, slowness to respond or move, and tiredness that feels overwhelming. If you notice someone showing these signs, help them stay warm with dry clothes or blankets, shield them from wind, give them a warm (not hot) drink if they are alert and able to swallow, and seek medical help if symptoms persist or worsen. In cold environments, planning ahead with an extra layer, a space blanket, or a lightweight emergency shelter is a simple way to reduce risk.

Contingency planning means thinking ahead about what you will do if your plans go wrong. It starts with telling someone where you'll be and when you expect to return. It includes carrying a basic first aid kit, a whistle, a flashlight, a small amount of food, and a device to contact help if you have one. It also means choosing a safe fallback plan in case conditions change or you run out of resources. Ask yourself: What will I do if a storm rolls in? What is my plan if I get separated from my group? What is my next best option if I cannot start a fire or cook with certain fuels? With a solid plan, you reduce the chance of getting stranded or caught unprepared.

Two micro-cases clarify emergency planning. Case A: You're in a high-wind area with a small group. You agree on a meeting point, carry a whistle, and designate a person to stay near the trailhead if anyone gets separated. You also plan to signal for help if you cannot reach your group within a certain time. Case B: You are on a cold night and rain begins to fall. You've packed extra dry clothing and a compact emergency blanket, you locate a sheltered spot, and you prepare a signaling plan to let someone know you're safe or need help.

Exercises you can do now (timed and measurable)

1) Fire rules check (15 minutes). Look up the fire rules for your current location or nearest park. If you can't find official guidance, ask a park ranger or a local authority. Outcome: You know whether lighting a fire is allowed and under what conditions; you can explain the reason to someone else.

2) Leave-No-Trace practice (20 minutes). On a short walk, pick up any litter you find and plan how you would carry it out. Practice using an established fire ring if allowed; otherwise, practice a no-fire alternative. Outcome: You can demonstrate a practical approach to keeping the area clean and minimizing impact.

3) Emergency plan drafting (25 minutes). Create a simple emergency plan for a typical outdoor outing: who to tell, what to carry, how to signal, and where to meet. Practice a 60-second explanation of your plan to a friend or family member. Outcome: You can communicate a basic plan clearly and concisely.

4) Hypothermia awareness drill (15 minutes). Learn the key signs in a short handout or sheet. Practice recognizing signs on a partner (without touching or medical assessment). Discuss how you would prevent hypothermia with proper layering and shelter. Outcome: You can spot warning signs and explain prevention steps.

Troubleshooting—common problems and quick fixes

- Problem: You're sure a fire is allowed but conditions look marginal. Quick fix: switch to a stove or no-fire method; reassess regularly with updated weather information.

- Problem: You light a fire and a stray ember sparks. Quick fix: immediately check the ring or pit, reduce or extinguish the flame, and extinguish completely if conditions worsen.

- Problem: You cannot extinguish a fire fully. Quick fix: retreat to a safe place, call for help if necessary, and make sure you have a plan for escape and communication.

- Problem: You’re unsure about local regulations. Quick fix: contact the local park service or ranger office, or check official signage; when in doubt, don’t light a fire.

- Problem: You forgot to pack out ashes or litter. Quick fix: return to the site as soon as possible to retrieve the waste and properly dispose of it in a designated container or area.

Tools and templates you can carry or copy

- Emergency Plan Card: list of emergency numbers, meeting point, and personal contact details; space for two friends’ contact numbers and your plan.

- Leave-No-Trace Quick Guide: space for checking burn status, ring or stove status, fuel limits, extinguishing steps, and trash pack-out plan.

- Signaling Kit Checklist: whistle, small mirror, bright fabric or tape, flashlight or headlamp, and a planned signal method.

Key Takeaway: Respecting rules, protecting nature, and being ready to act calmly in an emergency keeps you safe and helps others enjoy the outdoors too.

Transition: In the next chapter, you'll learn how to evaluate environments and pick fire-starting methods that fit the rules and the conditions, so you can stay prepared without compromising safety or the landscape.

Dear Reader,

Thank you for exploring this series focused on failures and the power of comeback stories. If you find value in these stories and believe in the importance of embracing setbacks as part of growth, we appreciate your support. Please consider sharing a positive review on the platform where you accessed this series.

Your feedback helps spread the message, inspiring more young readers to see challenges as opportunities and to believe in their potential. We understand that mistakes happen, and we thank you for your understanding.

www.ingramcontent.com/pod-product-compliance
Lightning Source LLC
LaVergne TN
LVHW020234210926
848481LV00016B/1957
9798295528194